LIGHTS OUT IN CHERNOBYL:

Account of a Nuclear Meltdown

Dr. David A. Noever

DEDICATION

To those innocents who lost their
lives and future health to the tragic
events of the Chernobyl meltdown.

CONTENTS

ACKNOWLEDGMENTS

This account was made possible in part by the Oxford University Department of Theoretical Physics and the generosity of the Rhodes' Trust.

1 PREFACE

"According to data in the ...possession of the KGB of the USSR, design deviations and violations of construction at Chernobyl AES...could lead to mishaps and accidents"

Top secret KGB memo signed by then Soviet Chairman Andropov, dated February 21, 1979, seven years before the Chernobyl accident

It took years to uncover. But the release of top secret Soviet archives has raised once again the hanging specter of what some observers have called the most horrific industrial catastrophe of the twentieth century. The case for this historical reading is not unsupportable: Chernobyl, nuclear meltdown at the V.I. Lenin Nuclear Plant.

On April 28, 1986, the explosion of Reactor Number Four dwarfed the bombs dropped a quarter century earlier at Hiroshima and Nagasaki. Chernobyl's shadow looms larger than those nuclear events that ended humanity's bloodiest war. When the thinnest of Chernobyl's nuclear reactor casings cracked, it pumped into the atmosphere thirty to forty times more radiation than both bombs put together. That hot spray continued burning unabated for three days, creating a radiating cloud which eventually circled the globe. The Swedish saw it first, then the wind reversed and easterly winds shrouded most of the northern hemisphere. The lifetime of that radiation is gauged not in days but in epochs. Its legacy will carry forward half-lives demarked in tens of thousands of years.

2 PEOPLE PLUS ELECTRIFICATION

In 1917 it was a straightforward, but powerful motto first chanted from speaker's platforms when the Soviet Union was born. Nearly to the day, seventy years later, electrification meant splitting atoms and the drive for centralized power production no longer heralded coal-black ovens, but a tight core of glowing nuclear rods.

The political timing of Chernobyl could not have sparked a

more heated internal Soviet debate: the end of the Cold War, *Perestroika*, liberalization of the press, the opening of foreign immigration--all can only be understood as incidents related to a singular nuclear blast breaking forth from the darkness at 1:21 A.M. in the morning.

It is therefore remarkable and steeped with irony that seven years before Reactor 4 exploded it was not *Pravda* nor any foreign embassy, but the KGB itself that warned about the potential Chernobyl danger in a series of top secret revelations to the Soviet Central Committee. As these confidential memos show, the Electrification board supposedly took corrective measures, so all could remain on schedule until one low-level subordinate launched a Soviet odyssey decidedly off-schedule.

These events encompass global issues, but the final toll of Chernobyl is profoundly human. No one can predict what will finally be the exact number of victims. Thirty-one lives were lost immediately. Hundreds of thousands of Ukrainians, Russians, and Belarusians had to abandon entire cities and settlements within the thirty-kilometer zone of extreme contamination. Estimates vary, but it is likely that some 3 million people, more than 2 million in Belarus' alone, are still living in contaminated areas. The city of Chernobyl is still inhabited by almost 10,000 people. Billions of rubles have been spent, and billions more will be needed to relocate communities and decontaminate the rich farmland. Upon reflection, the catastrophe launched a vital environmental 'green' movement in the former Soviet Union, but by some accounts, simultaneously brought down most of its coordinated empire. Even today, in the nearly bankrupt Ukraine, 15% of the total national budget goes to combat the aftermath of health bills and salaries for police and safety technicians.

The present account of the Chernobyl nuclear accident is based on the actual events as they unfolded. Interviews were conducted with the ordinary Ukrainian citizens who survived the blast in surrounding cities. The releases from secret KGB communications are now archived in Moscow. As part of 300 scholarly documents retrieved from a travelling exhibit at the Library of Congress, their contents have been translated admirably and called: *"Revelations from the Russian Archives."*

3 CITY OF CHESTNUTS

The Place: Kiev, Ukraine: The breadbasket of Soviet agriculture, an industrial heartland

The Time: April 16, 1987: Nearing the one-year anniversary of the first reports of a nuclear incident

"Imagine yourself walking down the street in a country where no one knows your name." The speaker introduces himself as Boris. He is a Soviet scientist invited to the International Conference of Physics convened in Kiev. "You're walking in Kiev, the Ukraine, where the women stop and smile."

Kiev was a beautiful city, heralded as the city of chestnuts. Its sturdy trees bear the meat of the nut which always, at least before this year, would ripen by early spring. In most years, the chestnut harvest announced the arrival of unpredictable skies and March, the lamb-and-lion month. White chestnut meat, already put in baskets and cracked by April, was sweet for the roasting just after the snow melted. Not this year. This year, the snow blanketed the parks and cobblestone streets, covering them from November until well into April, when previous springs had already flowered into summer.

No one knows why the snow fell harder this winter, but a few townspeople speculate. The summer before, radiation recorded in Kiev rose higher than normal, and reached higher in autumn too, until finally, more than nine months after a lone nuclear puff rose over their skies, the snow fell and cleared the Kiev air.

Soviet scientists, mostly physicists and meteorologists, report to the people of Kiev: the radiation on the streets is nominal. The snow has removed all but the most minute background exposure. Your lives should go back as they were before— before Reactor Four cracked and released more radiation than 30 Hiroshima nuclear bombs across the Soviet countryside.

But if the radiation no longer hangs unseen in the air, then it must have accumulated in the snow, or lie thawing in the slush that leaks into the soil. And in April, that snow, the same blanket that once cleaned the air, still covers the ground. The grass remains dormant. The limbs shine with hard brown dew, their buds encased in an ice coffin called spring frost. And the chestnuts have yet to drop.

In Kiev, a few speculate that this spring the snow stays because of a minor oversight left untended by an engineer, a lower subordinate who, while working the midnight shift, ran a battery of tests planned for nuclear reactor number four. Few in Kiev knew of that reactor's daily hum as it wired power from a little talked-about farming town. But now everyone knows the whereabouts of that community located thirty miles north of Kiev, the place called Chernobyl.

But it is Soviet Georgia not Chernobyl that is home to Boris. He is a physicist invited to join in the Kiev conference, a congress that will address the alternatives proposed to replace the sort of nuclear plant that exploded one year earlier on April 26, 1986.

That morning, the eruption shot a plume of flames that rose hundreds of feet, lighting the night air and scattering white-hot debris. And when finally subdued, the dying fire continued to fume and glow for three days more, spraying the countryside with six to seven tons of radioactive fuel. Along the nearby Prypyat River, unknowing residents went on sunning themselves, applying tanning lotion, thinking only of ultraviolet solar protection, and not glancing eastward towards the invisible meltdown of a nuclear reactor less than walking distance from their evacuation zone. Such a nuclear accident, one exposing the hot core, should happen once in a million years.

4 BORIS

Moscow, Russia, March 16, 1979: Memo stamped "secret" from the Deputy Minister of Power and Electrification, Comrade P.P. Falaleev, dated seven years before the Chernobyl accident.

Secret   Copy no. 1

USSR   MINISTRY   OF POWER AND ELECTRIFICATION (MINENERGO USSR)   Central Committee CPSU, Kitaiskii pr. 7 Moscow,   K-74   103074,

Minenergo USSR Central Committee of the CPSU Moscow K-ll A.T. 112604 MAR. 16, 79 07738 3-16-

79 No. 1381-2c; Return to General from Section of the Central Committee Checking the structural condition of the first unit of the Chernobyl AES.

The Commission found that violations of construction technology and design deviations have been documented.

At the present time the power station is operating successfully, and there are no hindrances to its further operation.

A letter of similar content was sent to the Central Committee of the Communist Party of Ukraine Deputy Minister.

[signed] *P. P. Falaleev,*
Kiev, Ukraine, April 16, 1987

Boris shows no signs of worry. He lives in Georgia, not the American state, but the Soviet republic, although during the conference of physicists, he will raise his glass of straight-up vodka to propose several toasts to "America, a great country", and to "Soviet Georgia, the fifty-first state of America, a land of soulful people." After hearing the blues song, Georgia on My Mind, a rolling classic, cut with the timbre of Ray Charles, Boris claims to understand the street-meaning of the word, "soul".

When describing the soul of Georgia, he winks: "God gave the Englishmen England, the Frenchmen France, and the Spaniards, Spain. And yes, the Russians, He gave

Russia. Then He spotted a man resting under a tree. It was a lone Georgian drinking cognac. This infuriated God. He commanded: "Thou must give up thy evil ways! Thou must never sin like this again!" The Georgian trembled with fear. "So always share, God commanded, if there ever is enough cognac for two."

Boris is a sharing comrade. When he found two tickets to attend the Soviet circus, he proposed to find, too, some young companion to host an unplanned visit to their world-renowned spectacle, the Soviet version of Ringling Brothers, Barnum and Bailey. He introduces his young friend as Olga, "practically an American" girl. And although the day before Olga had talked with me while touring Kiev, she suddenly looked nervous while riding in a cab headed for the circus. Twice she asked me to roll down the window, once to calm her nerves and then to cool her hot flashes. "This girl will take you," says Boris, the one responsible for planning the success of this circus excursion and hairless, appearing almost infantile in the face, even lacking eyebrows.

To get front-row seats at the last moment, Boris had had to call the circus director who, after hearing that a foreign scientist wanted to attend, arranged for two seats previously ticketed to a sold-out show. The director took the two tickets saved for Boris's parents. "And what does our foreigner think of Soviet efficiency?" asked Boris, as he held the tickets proudly, the two taken from his parents. "Nothing will be too good for our visitor."

Although married with a seven-year old boy, Olga has dressed in

Soviet smartness for the circus. "To compare to the women pictured in Vogue," she says. Olga wears a tartan skirt cut below her knees, but hemmed above vinyl boots. The pages of her Vogue would now be decidedly yellow with age. Like most women seen in Kiev, Olga has her hair chopped into a bob, a style that, according to Ukrainian custom, signifies and separates a married woman as unavailable and distinct from the "virgins still looking".

When we step from the taxi, Boris speaks in rough-house Russian, warning Olga in paternal tones. Later, after the circus lights have dimmed and the performance has begun, she tells me: "Boris said for me to follow the rules."

But it was exactly the rules that engineers at the ill-fated unit No. 4 were following. Just past midnight, when their circadian clocks called for strong stimulants, the engineers ordered a test of their emergency system and powered down Chernobyl below 25% capacity. It was a fatal decision. Two of those operators died. Among those who survived the initial fire wall, five were jailed. During their trial, the surviving men could not be found criminally at a loss. No engineering manual warned that a low-power test would automatically trigger a core meltdown. Only seconds before the fatal reactor instability began, did the operators see the floor tiles start to dance up and down.

5 SOVIET CIRCUS

Moscow, Russia, March 23, 1979:

Memo from V. Frolov, Head of the Section of Machine Building of the Central Committee of the CPSU, in answer to KGB safety inquiries from Chairman Andropov.

Secret

CENTRAL COMMITTEE OF THE CPSU Defects in the Construction of the Chernobyl

AES Chairman of the KGB of the USSR, Comrade IU. V. Andropov, reports about the low quality of construction work on various sections of the second unit of the Chernobyl AES of the USSR Ministry of Power and Electrification.

The specialists conclude that at present there are no obstacles to the subsequent operation of the Chernobyl AES, and the electric power station is operating reliably.

[SIGNATURE] (V. Frolov) March 23,1979 Nos. 05363,07738, A/H.

Kiev, Ukraine, the night of April 16, 1987

The Soviet circus, "the best anywhere," plays in an arena seating three-hundred, a chattering and clapping audience mostly made up of children attending with their parents.

"I have thought much about Chernobyl," says Olga. "Let my heart get light and we will speak at intermission"-- midway between the acrobatics and magic, the two halves of the world-respected Soviet circus.

Then the children go quiet. A drum pounds and rolls from behind stage, the curtain rises, the lights blacken, all except for a pen spotlight that dances over the crowd. To shatter the darkness, a magician lights a match, sparking a single flame, his white glove holding it, burning high in the air until the drum fades and the audience, formerly a din of gossip and fidgeting, now silences itself.

In tramp the players. Acrobats bounce and spin and tumble, whether balancing on a trampoline or trapeze, the latter suspended to a height rising ten stories. As practiced in Russia, tradition dictates that if a performer makes a mistake, if he stumbles or falls, he must repeat the same act, trying again and again if required, until the audience applauds its approval. "Soviet pride," explains Olga.

And fail, they do, much to the delight of open-armed, wide-eyed children. Parrots plumed in white, red and yellow fly mistakenly off-course into the crowd. Strong-men stagger and drop somersaulting midgets. Toe-dancing ballerinas trip in high heels. On command, dogs ride pigmy ponies half-way round the ring, until the trained canines spot a cat cradled in the audience. The hungry dogs return to the ring only when persuaded by the threat of a fifteen-foot whip, cracking and snapping its disapproval. All this folly brings titters and chuckles spilling from the audience. Olga, notably included, pours forth a jolly bundle of laughs.

But in the circus, pride and repeated performance play their part. It is not practiced success alone or perfection that gets the most enthusiastic applause. Rather it is failure followed by final success that meets loudest approval, louder even than a flawless act, if only because failure can first demonstrate the difficulty of the feat.

6 CLOWN FACES

Moscow, Russia, March 23, 1979:

Secret Memo from V. Frolov, Head of the Section of Machine Building of the
Central Committee of the CPSU, in answer to KGB safety inquiries from Chairman Andropov.
Secret
CENTRAL COMMITTEE OF THE CPSU

It was found that in erecting individual pillars, wall panels, and crane track stopways, there indeed were instances of deviation from the reference axes and substandard construction practices. Most of these deviations were exposed by the design inspectorate of the Gidroproekt Institute and a technical site inspection, and they were documented.

[SIGNATURE] (V. Frolov) March 23, 1979 Nos. 05363, 07738, A/H.

Kiev, Ukraine, the night of April 16, 1987

"We do not worry about Chernobyl," says Olga, beginning her story during the circus intermission. "If any circus feat appears difficult," she says, "then any clown's humor only become easier." She reasons that the serious players expend so much energy to defy nature, to climb higher or flip longer, that when a country bumpkin, a clown or fool, can admit failure, even cultivate the kick planted in the pants by boat-sized boots, then he brings the house down. To illustrate her point and to conclude the first act, out had stepped two clowns, one of which, if judged before by his winking mien tipped towards the audience, will be the inevitable winner and champion. The communist clown holds a purse of money and a string. And tiptoeing into the arena, his clowning comrade proceeds to chase the purse of money round the ring. He, the victim of a capital practical joke, never notices the string nor, peppering all around him, the hot cackles sounded by schoolchildren.

If tragedy can be gauged by this depth of humor evoked to foil it, then Chernobyl was a scream. "The plant had to explode." Olga's final confession is still buoyed with fleeting laughter of clowns. "Because if the power plant had not melted, we would have had electric lights to read by and therefore..." She raises her forefinger flamboyantly to make the point, "we would have been able to discover in our Pravda newspaper who was responsible for the mysterious disappearance of all our comrades."

7 OLGA

Moscow, Russia, March 23, 1979:

Memo from V. Frolov, Head of the Section of Machine Building of the Central Committee of the CPSU, in answer to KGB safety inquiries from Chairman Andropov.

Secret

CENTRAL COMMITTEE OF THE CPSU The Chernobyl AES construction directorate along with the design organization and client have adopted coordinated engineering solutions ensuring design reliability and construction quality. A timetable under the control of the Ministry leadership has been established to eliminate the noted deficiencies.

[SIGNATURE] (V. Frolov) March 23,1979 Nos. 05363,07738, A/H.

Kiev, Ukraine, the night of April 16, 1987

Olga walks through the meandering children waiting now for the second act to begin. The kids cluster around pictures showing the most famous circus performers in one of the world's most unique labor markets. The entire Soviet circus is said to employ only five families, accepting no one from outside this cloister of elite families. No child is admitted who has not been segregated from the community and taught, beginning at an early age, how to juggle and eat fire. To run away from home in the Soviet Union, a child cannot—as is customary in the West-run to the circus. Instead, if seeking new adventures, a Soviet child must run not to— but away from— the circus. Because to have ever performed here, your home always must have been the ring, the circus where you were born.

"I was in Kiev when the nuclear plant blew up," says Olga, gesturing for me to leave the arena to walk in the night air and to find a place alone. Was she evacuated? Olga looks around. "I should not say."

We walk past a boy unsuccessfully chasing a feather. Posted to the left, hanging on the door that frames a ball-bearing factory, we pass the international sign for 'not allowed here', a circle with a slash through the middle. And inside the circle is painted a red trumpet: "No horns blown."

"When the plant exploded, we knew nothing," begins Olga, secure that, finally, we are walking far from the circus.

She explains the explosion. A shell made of concrete surrounds most nuclear plants. This hard umbrella prevents radiation escaping from the hot core of the reactor and, in case of rupture, from leaking into the air. At Chernobyl, such a shell stood as safe cover. But it failed, shattering during the initial explosion, ripped into pieces and scattered among the cows lazily grazing nearby. The protective shell was built to survive only mild forces. At Chernobyl, the concrete walls were capable of withstanding less than twice the pressure of a bicycle's tire.

"A neighbor's husband-he is a driver-was called to Chernobyl. He had to drive away the first victims. That morning, he came to my room. He told me, 'Olga, some kind of a terrible thing has happened. Close your windows and do not go outside.' By afternoon, the shops were boarded up and the streets were cleared. Kiev is a city of three-million. There was not a soul outside."

In a country without an inquiring press, this story of a neighbor was not to be dismissed as a rumor. Western intelligence follows this line, showing respect for word-of-mouth information, at least when spying in the Soviet Union. The Central Intelligence Agency admits as much when they employ Americans working in the Moscow embassy. One of their job descriptions entails telling tales and small white fables. They begin their working day with a false rumor started in Moscow. The remainder of their day is spent collecting gossip from the streets and, by comparison, seeing how the original story changed, distorted or twisted. And Muscovites, nine million of them, spread rumors like they're viral.

"So I called an acquaintance of mine," continues Olga. "Her husband is a physicist. I asked him: has the nuclear plant blown up? The physicist laughed and laughed. He said, 'An accident? It is not possible!'"

8 MIKHAIL

Moscow, February 21, 1979: Secret memo from the Chairman of the Committee *[KGB] IU. Andropov*

Central Committee of the CPSU

USSR COMMITTEE OF STATE SECURITY [KGB]

February 21, 1979 No. 346 Moscow

Deputy Head of the Construction Directorate, Comrade V. T. Gora, gave instructions for backfilling the foundation in many places where vertical waterproofing was damaged. Damage to the waterproofing can lead to ground water seepage into the station and radioactive contamination of the environment.

Chairman of the Committee [KGB] [signed] IU. Andropov

Kiev, Ukraine, the night of April 16, 1987

At Chernobyl, it was the firemen, not the physicists, who arrived first. They came because fire lit the morning sky. The corpses of those first firemen were unidentifiable. They weren't burnt or scarred, but hot, their bodies contaminated by radioactive caesium-137. No doctor, at least not without risking radiation sickness himself, could get close enough to check the remains against existing dental records. Each firefighter, forever untouchable, received his burial sealed in a lead casket. Their funeral sarcophagi will rival the longevity of those built for the pharaohs. These first victims of Chernobyl were laid to rest in tombs-anonymous and unsafe to open for the next ten thousand years.

"I called the police," says Olga. "Has the nuclear plant blown up? The authorities reported no accident. But they advised everyone to stay inside.

"My seven-year old son, Mikhail, he was locked in the house all day. It was Sunday and his grandfather had come to visit. On weekends, he takes us riding in the country. Kiev, it was a beautiful city, always green and lush in the spring, and Mikhail loved to tear through the parks. We would collect sweet chestnuts and eat them from the shell until daylight disappeared.

"But that day, Mikhail stood at the window all afternoon. He must have known something was wrong, just judging by the looks frozen on our faces. It is terrible to wait like that, not knowing what will happen next, only knowing that, when it finally does happen, it will be what among all your fears, you very most fear".

9 HOT MIST

Moscow, Russia, March 23,1979:

Secret Memo from V. Frolov, Head of the Section of Machine Building of the

Central Committee of the CPSU, in answer to KGB safety inquiries from Chairman Andropov.

Secret

CENTRAL COMMITTEE OF THE CPSU To raise the quality of construction and assembly work and strengthen control over fire and radiation safety, the Ministry has published instructions strengthening fire-fighting procedures.

[SIGNATURE] (V. Frolov) March 23,1979 Nos. 05363,07738, A/II.

Kiev, Ukraine, April 16, 1987

In 1917, the Soviet Union banded together at a time when electricity was a modern miracle, glowing in lights that previously flickered, warming ovens that before burned black with charcoal, and promising cities and towns that sparkled and ran smoothly, like miniature electric trains. But scores of five-year plans later, the towns surrounding Chernobyl sparkled for a different reason and the people (not the ovens and lights) ran wildly, clearing the streets before the hot mist, a radioactive cloud, passed overhead.

"That night, Mikhail was still staring motionless at the window. Only seven years old. He had not said a word all day." Olga shuts her eyes, as if painting on the black canvases draped over her inner eyelids. She was reliving every moment from a night that began like many others, dated more than one year before.

"Finally, he climbs into his mother's arms, he is crying, and he says: 'We people, we can stay inside. But mother, what about the trees?'

"Just a baby." Olga's voice trails away as she wipes a bubbling tear from her cheek. "I am sorry," she says.

10 DOCTOR RICHARDSON

Moscow,   February 21,1979: Secret memo from the Chairman of the Committee [KGB] IU. Andropov

Central Committee of the CPSU

USSR   COMMITTEE   OF STATE SECURITY [KGB]

February 21,1979 No. 346 Moscow

The structural pillars of the generator room were erected with a deviation. Wall panels have been installed with a deviation. The placement of roof plates does not conform to the designer's specifications.

Chairman of the Committee [KGB] [signed] IU. Andropov

Kiev, Ukraine, April 16, 1987

An American scientist, Dr. Richardson, had befriended Olga while attending the Kiev conference. Richardson fashioned himself an eccentric. His beard grows stringy and untrimmed. His shirt shows stains splashed from his last meal and dried on from all the other meals eaten earlier in the week. And his eyes, colored pink and heavily veined, declare his dedication committed to long hours of work. Among scientists, his eccentricity is a cliché.

"Dr. Richardson told me," continues Olga. "He said, the public is paranoid about nuclear energy. Because they don't understand it. But not scientists, he said. We work around radiation all the time. And we seem no worse than everyone else, do we?"

By the end of the conference, Olga will never leave the side of Dr. Richardson. Together, they translate scientific articles, sparring back and forth while switching rapidly between English and Russian. They share dinner and they dance, clasping arm in arm.

"He made me feel better," says Olga. "I mean, a physicist should know about radiation. And while visiting Kiev, he seemed so relaxed.

"But I started to think," she continued. "I was not questioning his expertise. He seemed very smart. But I wondered about his self-assurance. Would a fireman act the same? I mean, once a fireman knows that fire is just chemistry, nothing more than a reaction between oxygen and carbon, does that mean fewer burnt faces come to his memories?

"No!" she answer "But who knows whether Kiev is safe? There is a famous Russian article written about the fallout expected from nuclear tests. After fallout, it takes a long time before the land becomes safe again. And meanwhile, waiting in Kiev, sleep three million people."

11 DIRTY SNOW

Moscow, February 21,1979: Secret memo from the Chairman of the Committee [KGB] IU. Andropov

Central Committee of the CPSU

USSR COMMITTEE OF STATE SECURITY [KGB]

February 21,1979 No. 346 Moscow

Similar violations were permitted in other sections with the knowledge of Comrade V. T.

Gora and the head of the construction group, Comrade IU. L. Matveev.

Chairman of the Committee [KGB] [signed] IU. Andropov

Kiev, Ukraine, late on April 16, 1987

As Olga walks around the circus arena, making her way through the wooded parks and crowds awaiting the second half of the show, she suddenly becomes confused. She's uncertain whether her story should be told. Moreover, whether, speaking amidst the bustle of comrades, she should be the one talking. No one else in Kiev, at least no one except foreigners, will talk about Chernobyl. But confusion, and finally fist-swinging aggressiveness, are two of the symptoms linked to the painful disorder called "radiation sickness".

"Most of the fuel spilled at Chernobyl is gone," she says, repeating the official party line: no danger. "Eighty to ninety percent of the nuclear material has left the site."

So Kiev is safe?

"Oh yes," she says. "Most of the fuel went into the atmosphere."

Darkening at dusk, the sky flashes bits of electric colors, mostly sunset violets shining behind an otherwise gun-grey haze. Packed solidly below, the snow sits, boot-blackened and sooty, accumulated into hip-high drifts.

"Radiation recorded in Kiev was above average, high even, in early fall, more than six months after the explosion. But when the snow fell, and it fell, as you say, like dead cats and dogs, then the radiation levels found in the air dropped back to normal."

And how long has "this snow" accumulated on the ground? This snow or its grime and sleet equivalent has drawn the attention of two passing strollers. Just then a Russian schoolboy has packed a bundle of snow until, when pounded into a fist-sized ball, the ice melts enough to harden. He eats the frigid remnants of previous clouds and snowstorms like a white apple.

Standing behind the boy, a brandy-nosed drunkard weaves along the pavement, stopping twice to test the depth of the snow by digging in the point of his boot. Olga rolls her eyes at the drunk, as if to say: stupid drunk, please do not stumble in front of a foreigner. The drunkard's feet disappear to ankle-depth, both of his cuffs filling with drifting ice.

"Oh, this dirty snow?" says Olga casually. "Ever since the fall."

12 TRAINED CATS

Moscow, February 21,1979: Secret memo from the Chairman of the Committee [KGB], IU. Andropov

Secret

Central Committee of the CPSU USSR COMMITTEE OF STATE SECURITY [KGB]

February 21,1979 No. 346 Moscow

Construction Flaws at the Chernobyl Nuclear Power Plant [AES]

According to data in the possession of the KGB of the USSR, design deviations and violations of construction and assembly technology are occurring at various places in the construction of the second generating unit of the Chernobyl AES, and these could lead to mishaps and accidents.

The KGB of Ukraine has informed the CPSU Central Committee of these violations. This is for your information.

Chairman of the Committee [KGB] [signed] IU. Andropov

Kiev, Ukraine, April 16, 1987

If Olga takes the tour further, she will get lost and the second half of the circus will begin with two empty seats. "It is a crime what they did in Kiev," she says, pointing the way back to the circus. "It raises the hairs on my neck. The people in Moscow—, the politicians, the scientists, all of them— they committed two crimes." Olga curls her fingers into a fist. "First, they never warned us about the exploding nuclear plant. And second, the worst, they never warned us, not because they didn't know—because they did—but because if we knew, if the public really knew the truth, then we would become outraged." Clasped to her breast, Olga holds her two fists, balled so the blood drains. "Outrage, ha! I will tell you something about terror. "

"But here," interrupts Olga, "we will be late for the magic show." At the circus, the sudden appearance on stage of the illusionist, Keo, a 'marvel of the supernatural', brings rowdy applause, most enthusiastically from Olga. During the first act, she had feared the entrance of performing animals, her particular circus terror, despite the reassurances otherwise of the show's director. In the interest of travelling economy, the Soviet circus uses human performers mostly and only occasionally asks for a trick from an animal or two. The director did say, however, that tonight a cat might perform, but otherwise Olga need not worry about wild or stray beasts. At the time she took 'cat' to mean a real lion or tiger, not domesticated cats, the tabby or Siamese that, in truth, are much harder to train and tantalize than their wild cousins. In fact only one Soviet clown, the only such trainer in the world so gifted, can coax, lure and ultimately command domesticated cats to perform feats elsewhere considered beyond the teaching prowess of even the most accomplished animal trainers. So here they came, ordinary house cats performing gymnastic jigs, usually headstands and rollovers. This clown does not bait his cats. But when the spotlight glares, the clown performs the impossible, a feat as inexplicable as it is rare, the fickle feline that does exactly as it is told.

To conclude following these most cunning of felines, it is the show's magician, Keo, who announces his entrance and the evening's finale. In his black cape, he claps his hands and, before on-looking eyes, the seemingly impossible begins to happen. From the audience, Soviet women begin to disappear. The trick appears simple: again and again put girls in boxes, urns, more boxes in boxes, cages, and paper cylinders. Then make the women disappear, shove them through glass, shoot, burn and saw them in half, turn them into doves and lions, Olga's most singular fear, until finally the underfed females step into the spotlight, their toned forms beaming and bowing to the music of applause.

And those women, showing more than peasant faces, smile through the torture. Keo favors one Soviet type: svelte and lithe in motion, lean-boned like Audrey Hepburn, but marked, distinct from her Western counterparts, by an eagle-nose.

On first appearance, this feminine ideal, never seen on the streets of Kiev, brings "oohs" and "aahs" from the audience. But as Olga acknowledges with a pointing finger, their womanly beauty is more than mere skin. The magician's helpers wear bikinis, little more than G-strings cut with jewels, their costumes dripping pearls and ruby glass, their get-ups quite exposing by Soviet standards and seemingly ice-cold. "Their dresses are beautiful," says Olga, her eyes widening at their lack of dresses at all.

But the magician's trick, the disappearance of these earthly angels, reveals itself as even simpler. The floor of the arena is plastic and etched with trap doors. In the spirit of fun, the children cackle and Olga shrugs a sigh of contentment.

13 RUMORED TRUTHS

Moscow, February 21, 1979: Secret memo from the Chairman of the Committee [KGB] IU. Andropov

Central Committee of the CPSU

USSR COMMITTEE OF STATE SECURITY [KGB]

February 21, 1979 No. 346 Moscow

The leadership of the Directorate is not devoting proper attention to the foundation, on which the quality of the construction largely depends. The cement plant operates erratically, and its output is of poor quality. Interruptions were permitted during the pouring of especially heavy concrete causing gaps and layering in the foundation.

Chairman of the Committee [KGB] [signed] IU. Andropov

Kiev, Ukraine, late on April 16, 1987

And when the last circus performer's bow rises, Olga elbows me. She rushes past my seat, her run headed for an exit, her hands held to her mouth, as if suddenly forgetful or mysteriously sickened. Olga was forgetful. She wanted to warn me. The cloak room operates on the communist rule: first-come, first-served. And in the mob of departing attendees, the last-served must wait and wait, one among many, in a line of hair-twirling, impatient children.

Olga turns to remark over her shoulder, "You may have heard the proverb: When a communist butcher slaughters a cow, who gets the filet mignon? We comrades all must share--even our waiting time."

Once in line, Olga is not impatient. For her one week of work, because she has translated for a foreigner, she will receive three months of holiday, a vacation far from the back-breaking work ethic of her comrades. In Olga's case, she will travel to Odessa, by the Black Sea, where her husband and she hope to "thaw the frost from our noses." Until then, Olga has nothing but time, hours to kill.

"After the explosion at Chernobyl," she begins, "No one went onto the streets. Not for a few days at least. And not because we were warned against anything, but simply on the truth of a rumor. Workers did not work. Students did not learn. The people risked losing their jobs," singularly the most talked-about possession even in a country without any recognized forms of unemployment, "on their faith in gossip. That is how frightened we were.

"So on the third day, I rang my physicist friend again. This time he was sober over the telephone, telling me a different story than before. Yes, there was an explosion, yes, it was nuclear, and yes, whatever you do, take your child and get him out of Kiev.

In Your Face IRS: Zero Taxes

I am not talking here how mega companies shift their revenue to off-shore tax havens. I am also not talking here how billionaires pay hardly 14%-15% of their income in taxes (14%-15% is too much for me). I am talking here how middle class America can pay zero taxes. If you are a farmer/rancher, mom-and-pop store owner, contractor, doctor/lawyer/accountant providing services, small business owner, you should not pay a dime in taxes.

Students don't even have any earning, may be $8-$10 an hour earning from a burger place; taxes are not really an issue for them yet, but ballooning unforgivable student loan is. What I am talking here is that they need not to pay it back.

No, don't hide your income; don't break any laws. I am in-fact suggesting to follow the law to the letter. IRS can't do a sh**. IRS doesn't make laws, they follow laws. Congress makes laws. After reading this book, will congress change laws I am referring to? I challenge them, no, I dare them. If members of the congress even think to touch these laws, they will shoot in their own foot; their own assets will become venerable. .

Individual's act is small but it validates a hypothesis. Once we have proof-of-concept, we can repeat it – basic scientific principle. I would like to see that all middle class Americans eliminate their taxes.

Rochit Rajsuman

A cool Silicon Valley nerd, Rochit Rajsuman received Ph. D. in Electrical Engineering from Colorado State University. He has worked both in the academia and industry in engineering and engineering management positions. In academic career, Dr. Rajsuman has served as Assistant Professor in Computer Engineering and Science Department at Case Western Reserve University and as Pinson Chair Professor in the Electrical Engineering Department at San Jose State University. In industrial career, he has worked at Silicon Valley semiconductor companies LSI Logic and Equator Technologies and as manager of research at Advantest America R&D Center and as Chief Scientist at Advantest America Corporation. He has also founded two technology companies.

Dr. Rajsuman has authored three computer/electrical engineering books, all published by the Artech House Inc., numerous papers and 30 awarded US patents related to design and testing of semiconductor integrated circuits.

ISBN 9781477640456

90000 >

9 781477 640456

JustifyWorld Inc.®

"But in the city, we were safe for a few days. Nature has a fickle streak, like those circus cats. When Chernobyl exploded, the wind was blowing to the West." Olga laughs, her sausage-length fingers checking a nervous guffaw. "The life of my boy, my only son, depended on a breeze. Something as uncertain as the spin of a weather vane." Olga laughs again, this time to cover her caution. "The worst Soviet radiation, it went to the other countries, Poland and Sweden.

"But on April 30, the wind changed. And the cloud, this enormous hot mist, came toward Kiev. Three million people, all afraid to leave their homes for bread or open their windows for air. It was like a film about monsters, some invasion of enemies made of gas. You cannot fight them. You can only wait for them to arrive.

"The cloud was coming. My friends told me to take Mikhail to Moscow. My husband and his mother would take care of him there. Understand me, we still knew nothing from the authorities. They said: "*nyet* explosion, *nyet* danger." How do you say in English: 'all blue skies'. So by Mayday celebrations, Mikhail was safe, sleeping on a train that was steaming away from the coming hot snow.

"Then, I went home and burned every stitch of clothing I owned.

"Hot snow, that is what it was like. Beyond imagination. How do your American poets say, like 'whore'? In the morning, when the dew covers the ground?"

"Like hoarfrost," I say.

"Yes," smiles Olga. "Hot hoarfrost fell from the sky. Outside the city, it fell harder and looked like silver flakes, like the scraps from a shredded car, only as light as feathers, falling, almost floating in gusts until, by morning, the hoarfrost covered the farms and the factories and the trees, just like my Mikhail had predicted.

"In the trees. You know, radiation concentrates in some trees and fruits but not in others. See there," Olga points," the aspen trees, they are supposed to be safe. But the lime trees and the chestnuts, they are dangerous. Kiev was the city of chestnuts. But not this year. The chestnuts are hot this year and my Mikhail, he will stay in Moscow until the sweet meat of the chestnuts is safe to eat again.

"You'll never guess what they plan to do at Chernobyl?

'This is gossip, so I trust it. The scientists do not want to cut down the trees and bury them like the reactor core. They say clear-cutting would release more dust, some sort of radioactive ash into the air. Instead, they have invented another machine, a gadget, so they claim, that hammers the trees into the earth. Literally knocks them like pegs into the ground. Then, they will plant new seedlings which as they grow, will absorb the radiation from the soil.

But stuffing trees back into the soil, wouldn't that bring another generation of radioactive seedlings? Olga shrugs: "It all sounds so absurd. I mean, who would think to drive down a sixty-foot tree, bam, like a carpenter's nail?

"But then, who would have predicted that you would need to?

"I should not fret," says Olga, her voice a whisper. "Dr. Richardson told me, many cities are worse off than Kiev. And besides, before the cloud dispersed, it travelled at least once around the world.

"So even if Kiev is not safe, where would be?"

14 HOARFROST

Moscow, February 21, 1979: Secret memo from the Chairman of the Committee [KGB] IU. Andropov

Central Committee of the CPSU

USSR COMMITTEE OF STATE SECURITY [KGB]

February 21,1979 No. 346 Moscow

Access roads to the Chernobyl AES are in urgent need of repair.

Chairman of the Committee [KGB] [signed] IU. Andropov

Kiev, Ukraine, late on April 16, 1987

Olga walks the Kiev streets at night.

"We import all our food now," Olga begins. "We live on bread from elsewhere in Russia. But how can we know that it is safe? The people in Moscow, they lied to us once. They said there was no explosion. The Scandinavians had to tell us—we are Russians and strong—and even then we did not know for certain until the streets were covered in silver snow. Hoarfrost.

"So now we get our oranges from Sicily and our rice from the East. Dairy products come from Europe, but only when there is refrigeration. Of course, Dr. Richardson told me, and he winked when he said this, that these

71

products, oranges and milk, are exactly where radiation concentrates. And he said—I could not believe this: But in Russia, radiation will concentrate in the bread."

The next day, Boris scoffs at the mention of Chernobyl: "They made a mistake, that is their problem." But a mistake is not what Olga had in mind, not when she burned all her clothes, and not when she saw the silvered flakes falling, like hot snow, hoarfrost. At the circus, Olga had loved one trick in particular, one that she couldn't stop mentioning as we walked home through the bustling, night-hustling crowd. The audience in Kiev had stood to its feet to applaud this piece of magic and illusion.

The clowns had begun again, this time laying down a tarpaulin to convince the audience that, with all the floor's trap doors covered, no escape was possible. And into the silenced arena strode a shocking beauty, a Soviet girl gamboling in her bikini of gems, pearls and rubies. For all to see, she steps onto a platform, then with a shimmy, pulls up a dressing screen fashioned of paper. Over the top comes her bikini. Flying earrings and her necklace follow. And with the girl safe and secure and nude, the audience must wait.

The magician fumbles and drops his book of matches. He plans to set fire to the dressing screen and, as he leads the audience to believe, to the sculptured female behind. When one match finally sparks, they all do, erupting a flame that lights the otherwise darkened arena.

At the sight of fire, itself flaming so close to the paper screen, the audience gasps and the girl, unprotected, mouths a scream to set hairs on end, a tingler. The scream lasts until the last of the screen burns into cinder, the audience seated as far back as the fifth row covering their faces against the singeing heat, a metal-red burn. The girl's string of plastic pearls melts into a bubbling pool from the intense, bone-searing temperature. The audience waits again, this time for a view of the inevitable melted and scarred remains to emerge, blackened like the showgirl's string of pasty pearls.

But out from the charcoal steps the girl, victorious and clad again, unscarred by the tonguing flame, safely rising from a previously unseen urn of water. And as she takes her final bow, above her, from the ceiling of the circus fall the grey ashes that were once her dressing screen. The ash fragments snow down, like hoarfrost, grey-light ashes blanketing the Kiev audience, soot covering their clothes, flakes brushing through their hair.

This time the audience stood to applaud. The children stamped their feet. Exuberantly they waited for the departing bow of Keo, the singular magician who had already disappeared. Their feet stomped louder, pounding worn soles, shoes impossible to repair or replace because, this year in Kiev, cow's leather will likely be very scarce.

The citizens limp on, cleaning the streets of radiation, sweeping with each step, dancing on worn soles.

Fifteen Chernobyl style reactors still operate in the nearly bankrupt districts of the Ukraine, Russia and Lithuania. In 1991, following a devastating fire, Reactor No. 2 was forced to shut down its uranium fuel rods. Within 200 feet of the radioactive wreckage of Reactor No. 4, Chernobyl Reactor No. 3 continues its operation today, hidden behind a concrete wall.

ABOUT THE AUTHOR

Dr. David Noever graduated from Princeton University, *summa cum laude*, and Oxford University as a Rhodes' Scholar with a Ph.D. in Theoretical Physics. He worked as a NASA space scientist at the George C. Marshall Spaceflight Center in Huntsville, Alabama. He traveled to Kiev near Chernobyl during the first year following the tragic accident at Reactor Number Four.